# IF DINOSAURS WERE HERE TODAY
# ANCIENT BEASTS

Copyright © 2024 Bright Bound Ltd

First published in 2024 by Hungry Tomato Ltd
F15, Old Bakery Studios, Blewetts Wharf, Malpas
Road, Truro, Cornwall,
TR1 1QH, UK.

A CIP catalogue record for this book is available from
the British Library.

ISBN 9781916598959

Printed in China

Discover more at
www.hungrytomato.com

Picture Credits:
(abbreviations: t=top, b=bottom, m=middle, l=left,
r=right, bg=background)

Alamy: pegaz 9br, 10-11bg. Getty Images: Science
faction 1bg, 7bl, 9bl, 18-19bg. Shutter-stock: 2l, 4bg,
7mr, 7tl, 8br, 8mr, 9tr, 12-13bg, 14-15bg, 16-17bg;
Adwo 31tl; Anton chalkov 26bl; Catmando 9ml, 22br,
23br, 23ml, 27br; Chromographs 31mr; David.costa.
art 6tl; Denis - - S 6mr, 9mr; Dm7 27tl; Dotted Yeti
22tl, 24mr, 29bl; Elenarts 27mr; Engineer studio
6bl; Herschel Hoffmeyer 28tl; Hydebrink 26tr; JMx
images 5m, 23tl; Kostiantyn Ivanyshen 8bl, 9ml;
Krugloff 30tr; LukasKrbec 27ml; Mark Brandon 30bl;
Melissamn 28mr; Noiel 29mr; Orla 24tl; Penny Hicks
31br; Vectorpocket 29tl.
Stevebloom 3m, 8tr, 20-21bg.

Every effort has been made to trace the copyright
holders, and we apologise in advance for any
unintentional omissions. We would be pleased to
insert the appropriate acknowledgements in any
subsequent edition of this publication.

# IF DINOSAURS WERE HERE TODAY!
## ANCIENT BEASTS

by John Allan
Illustrated by Simon Mendez

HUNGRY TOMATO.

**WARNING!**
These extinct beasts are not alive today. But just imagine if they were...

# CONTENTS

Words in **BOLD** can be found in the glossary.

# THE STORY OF THE DINOSAURS

Planet Earth is around 4.5 billion years old. Rocks containing traces of living things shows us that there's been life on Earth for around 3.6 billion years. During Earth's long history, the planet and the creatures that roam it have changed drastically. We've all heard of the dinosaurs, but where did they come from, and where are they now?

## WHEN DINOSAURS ROAMED

Dinosaurs were the most famous and fascinating animals to come from these prehistoric times. Dinosaurs were the biggest land-living creatures to have ever lived. Alongside these giants lived smaller, bird-like dinosaurs, flying reptiles and huge ocean-dwelling beasts.

Then, 65 million years ago, the dinosaurs were suddenly gone! Scientists believe that a huge asteroid hit Earth, wiping out most living things. The extinction of the dinosaurs allowed for the rise of new animals: 4 million years ago, humans appeared!

## FOSSIL FINDS

Humans began observing rocks and, in the 1700s, discovered that the **fossils** they contained were the remains of ancient plants and animals. Fossil hunting became popular and the study of fossils – **palaeontology** – was born.

In 1842, scientist Sir Richard Owen invented the term 'Dinosauria' to describe the giant creatures that had once walked the Earth. Their remains fascinated both scientists and ordinary people – everyone wanted to know what these creatures had been like.

## THEN AND NOW

For over two centuries, dinosaurs have amazed and fascinated us. We wonder how they'd compare to the animals of today.

Could prehistoric meat-eaters adapt their hunting style to the prey that's available today?

Could the ancient Ouranosaurus live alongside modern camels in today's deserts and scrubland?

Would prehistoric sea creatures like Cryptoclidus be big enough to fight off today's ocean predators?

## THE UNKNOWN

We may never know exactly what it would be like to live with dinosaurs. We can only imagine, taking what we've discovered from their fossilised remains, and comparing it to what we know about modern animals to picture what life would be like if dinosaurs were here today!

**If you've got the courage, read on...**
...be prepared for some truly bizarre and spine-tingling - though imaginary - encounters between human or animal and beast.

# TIMELINE

### ARCHAEOPTERYX

Name meaning 'ancient wing'.

This primitive creature was like a dinosaur as it had the same jaw, claws and tail, but like modern birds it also had feathers and well-formed flying wings.

### TRIASSIC PERIOD

### [252–201 MILLION YEARS AGO]

Dinosaurs appeared towards the end of the Triassic **period**. They tended to live by the seaside, along riverbanks and in desert **oases** where water was plentiful. Early dinosaurs included Plateosaurus and Coelophysis.

### CRYPTOCLIDUS

Name meaning 'hidden collar-bone'.

This huge swimming **plesiosaur** shared the oceans with ichythosaurs during the Age of Dinosaurs.

### JURASSIC PERIOD

### [201–145 MILLION YEARS AGO]

During the Jurassic period, Earth's climate became moister and milder – new plants and forests grew, meaning new food sources for plant-eating dinos. As a result, both plant- and meat-eating dinosaurs started to grow much bigger.

### SEISMOSAURUS

Name meaning 'earthquake lizard'.

With an elephantine body and very long neck and tail, this plant-eater was one of the biggest dinosaurs to ever walk the Earth.

## OURANOSAURUS

Name meaning 'brave monitor lizard'.

Ouranosaurus was a relatively large, heavy dinosaur. Palaeontologists think that it mostly walked on all four legs, but could run quite quickly on just its two strong hind legs - particularly if being hunted by a predator!

## CRETACEOUS PERIOD

### (145–66 MILLION YEARS AGO)

This was when some of the most famous dinosaurs lived, including T.rex, Triceratops and Spinosaurus. Who knows what other dinosaurs would have lived since then if they hadn't all been wiped out by the huge meteorite?

## MASS EXTINCTION

For millions of years, dinosaurs ruled the Earth, until there was a **mass extinction**. There is evidence that a **meteorite** struck Earth around 65 million years ago, exploding rock fragments, causing **tsunamis** and forest fires, resulting in the death of the dinosaurs and all other reptiles of the time.

## AMMONITE

Name comes from 'Ammon' – after the Egyptian god who had a pair of curled goat's horns.

There were thousands of species of this octopus-like creature that lived in coiled, patterned shells.

## MICRORAPTOR

Name meaning 'tiny hunter'.

Is it a bird or a dino? This **controversial** creature may be one of the smallest we've discovered, but with sharp claws on its talons, this meat-eater was fierce.

# CLINGING ON
## MICRORAPTOR

Spots? On the face of the Statue of Liberty? No, not spots, but a flock of the tiny flying Microraptors. With wing feathers on both the arms and the legs, and a small, lightweight body, Microraptor is thought to have been a very efficient glider. Its four wings use rising air currents to lift it to considerable heights, and its long claws allow it to cling to the faces of tall buildings.

In Early Cretaceous times, Microraptor was one of several half-bird, half-dinosaur animals. For years, palaeontologists were convinced that no dinosaur could fly. The discovery of Microraptor, and similar creatures, have caused much debate about this. Even now, some argue over whether Microraptor is in fact a dinosaur or a prehistoric bird. If we saw a flock of Microraptors swirling in the slanting sunlight, we would think that they were birds or even big butterflies. These ancient creatures fed on small birds, lizards and fish. There is debate about whether Microraptor was more of a hunter or **scavenger** due to its size, but either way, its sharp claws and talons would've been useful!

## MICRORAPTOR
### PRONOUNCED
mike-row-rap-ter

### LIVED
Early Cretaceous period
125-120 million years ago

### WINGSPAN
up to 1 metre (3.2ft)

### DIET
**Carnivore**

## TINY DINOSAUR

For 150 years we thought the Compsognathus was the smallest dinosaur known. That was before the discovery of Microraptor in 2000, in China, and this turned out to be so small and lightweight that it could actually glide from tree to tree.

## MYSTERY FEATHERS

It's rare for palaeontologists to find well-preserved feathers, however, they can study fossils' pigments to get a good idea what they were like. They've recently studied one Microraptor and discovered its feathers were a glossy blue-black. It's possible different species had different feathers - some might've been brighter, like tropical birds today.

# THE LONGEST DINOSAUR

## SEISMOSAURUS

Chaos at the airport! A herd of Seismosaurus has wandered onto a runway. All traffic is brought to a halt as the great beasts wander unconcerned among the aircraft. It will take the security and maintenance services some time to move these huge obstructions.

Open areas, such as airport runways, would be a familiar **habitat** to big **sauropods** like Seismosaurus. They lived on the **arid** open landscapes of the Late Jurassic period, feeding on the trees that grew along the sides of rivers. They wouldn't be disturbed by the movement of taxiing aircraft, as to them they'd be just like other big sauropods to which they are accustomed. They may be more surprised to see the fast-moving vehicles taking off into the sky. That's certainly not something these huge creatures could do!

### SEISMOSAURUS

**PRONOUNCED**
size-mow-saw-rus

**LIVED**
Late Jurassic period
159-144 million years ago

**LENGTH**
up to 37 metres (120ft)

**DIET**
**Herbivore**

### QUESTIONABLE FIND

When the single-known skeleton of Seismosaurus was found, it was celebrated as the longest dinosaur ever, with an estimated length of 45 metres (148ft). But the skeleton was incomplete. Today, we don't think that it was quite so long. In fact, after close examination, scientists re-classed the Seismosaurus as a new species of Diplodocus!

# KEEPING COOL
## OURANOSAURUS

An exhausted camel rests in the shade of an Ouranosaurus' sail. It's summer and temperatures in Egypt have soared! There aren't many places to cool down. It's not a problem for Ouranosaurus because of its sail. Supported by long spines and possibly filled with blood vessels, the sail can be held into the wind to cool the animal's circulating blood.

Some scientists think that the spines on Ouranosaurus' back supported a fatty hump rather than a sail. The lump may have been like that of a camel, and used as a source of nutrition when conditions were really harsh. With the increase in global temperatures, prehistoric animals with exotic heat regulating devices, like Ouranosaurus, might do well if it lived today.

### OURANOSAURUS
**PRONOUNCED**
or-ran-uh-saw-rus

**LIVED**
Early Cretaceous period
115-110 million years ago

**LENGTH**
up to 7 metres (23ft)

**DIET**
Herbivore

## LARGE SAIL

The Ouranosaurus' sail was supported by a series of long, broad spines down the backbone, arranged like a picket fence. It may look unusual to you, but this sail wasn't unique. The big meat-eater Spinosaurus had a similar structure. It may have been used for warming and cooling the body, it may have supported a fatty hump for food storage during hot seasons, or it may have had a different function entirely!

# CHILLING OUT
## CRYPTOCLIDUS

Tourists watch the sea lions on the decks of a marina. They even encourage them by throwing scraps of fish. But now the plesiosaur Cryptoclidus has learned of this, it's coming to the marina for an easy meal, too. It would not be surprising if they also became used to human presence and adopted habits that we usually associate with today's marine mammals.

Cryptoclidus was one of the long-necked plesiosaurs - the marine reptiles of the Age of Dinosaurs. They lived and hunted rather like modern sea lions, 'flying' through the water with their four paddle-like flippers, and snapping up fish with their long, interlocking teeth. Also like sea lions, they probably came ashore to breed. It's likely that Cryptoclidus laid eggs and buried them in the sand while the babies grew, similar to modern-day reptiles like turtles. If Cryptoclidus lived today, we may fence off their nesting sites to protect them from human interference.

## CRYPTOCLIDUS
**PRONOUNCED**
krip-toe-cly-dus

**LIVED**
Late Jurassic period
166-145 million years ago

**LENGTH**
4 metres (13ft)

**DIET**
**Piscivore**

### MASTER OF DISGUISE
Scientists believe that Cryptoclidus' small head and long neck helped it to sneak up on prey. Potential prey were not threatened by its small head. Its long neck kept its large body out of sight – until it was too late for them to realise they were dinner! Cryptoclidus seems to have mostly eaten squid and small fish.

### FILTER FEEDING
Cryptoclidus had many interlocking needle-like teeth that may have been used to filter its food and small prey from water or even sand – in a similar way to modern whales who use bristle-shaped plates, called **baleen**, in their mouths.

# HISTORIC SEASHELLS
## AMMONITE

A sea otter surfaces in its rocky cove. After a brief underwater hunt, it brings up an ammonite – a tentacled sea creature in a coiled shell. Used to dealing with hard-shelled animals like oysters and scallops, the otter will find little difficulty in breaking into the shell and extracting the flesh.

Imagine an octopus in a coiled shell - that's what an ammonite looked like. Ammonites were amongst the most common sea creatures during the Age of Dinosaurs. There were a vast range of shapes and sizes, with different lifestyles ranging from fast hunters to drifting filter feeders and scavengers. In modern oceans, they would probably be as plentiful as they were back then.

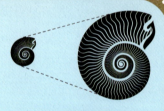

**AMMONITE**
PRONOUNCED
ah-mun-ite

LIVED
**Mesozoic** period
250-65 million years ago.

SIZE
Varied from less than
2.5cm (1in) up to 3 metres
(9ft)!

DIET
Piscivore

## MYSTERY SHELLS

Before palaeontologists understood what ammonites were, they called the fossil shells "snakestones" as they looked like coiled snakes made of stone!

## FLOATING SHELL

Ammonites are extinct today, but during the Age of Dinosaurs the seas were full of them. They mostly drifted in the water, adjusting their buoyancy by filling chambers in their shells with air. The picture (left) shows a cross-section of an ammonite with its chambers exposed.

# SMALL BUT FIERCE
## ARCHAEOPTERYX

A bald eagle, king of the skies, is flying back to its nest with a lizard. Suddenly it's set upon by a trio of arrogant Archaeopteryx. The eagle clutches its prize and tries to fend off the attack. It's not used to being challenged by other creatures.

For years, the Archaeopteryx was considered the earliest bird to exist, since it was covered in feathers and had wings that were almost identical in structure to those of modern birds. However, its body, jaws, claws, and tail were like that of a small dinosaur. This confused the palaeontologists studying it. Was it a bird or dinosaur? Whilst many now consider it a bird-like dinosaur - following the discovery of several other bird-like dinosaur species - there are strong arguments to be made either way.

## ARCHAEOPTERYX
### PRONOUNCED
ar-kee-op-ter-iks

### LIVED
Late Jurassic period
150 million years ago

### WINGSPAN
up to 60cm (2ft)

### DIET
Carnivore

## BIRD OR DINO?
If we found the fossil of Archaeopteryx without the feathers, we would think it was a little dinosaur. In fact, one specimen found was considered a dinosaur for years. Scientists believe this is strong evidence that birds are related to dinosaurs.

## CLOSE RELATIVE
Archaeopteryx's strange mixture of features would certainly look odd today, but its discovery is important as it suggests that birds and dinosaurs are closely related. But whether or not Archaeopteryx would survive today alongside highly-adapted modern-day birds is not clear.

# DINO DIETS

Dinosaurs had BIG appetites, but just like humans, they didn't all eat the same foods. Scientists have found fossilised dino bellies and poop containing different things, so we know for certain what some of them ate. Whilst we don't always have lots of evidence, scientists think they've worked out what lots of dinos ate for dinner.

## HUNGRY HERBIVORES

Most dinosaurs were herbivores – they only ate plants. You can think of them as prehistoric vegetarians. Earth had some different plants during the Age of Dinosaurs compared to today. There was a lot less grass, and more short-needled **conifers**, like cypresses and Araucaria trees for dinos to munch on.

Some plant-eating dinosaurs, like Seismosaurus, had long necks which helped them reach and eat from the tops of trees. Smaller herbivores feasted on low bushes and grasses. Most had flat teeth which helped them strip the leaves off plants and crush their food before swallowing. It's thought that dinos with weaker teeth swallowed stones to help grind up food inside their stomachs!

## PREDATORY CARNIVORES

It's thought that about 35% of all dinosaurs were carnivores, meaning they only ate meat. They had strong legs, jaws and claws which would help them to hunt, catch and eat their prey. Reptiles, eggs, and even other dinosaurs were on the menu!

Many of the big carnivorous dinosaurs, like T.rex, were at the very top of the food chain. This meant that nothing tried to hunt or eat them, making them top predators. We know from fossil evidence that they hunted live dinos as well as scavenging dead carcasses. All living things must have hidden when they heard a T.rex coming!

Although a lot of carnivorous dinosaurs were huge, terrifying predators, they weren't all big. One of the smallest meat-eating dinosaurs was the Microraptor who was not quite 1 metre (2.5ft) long! Don't be fooled by its size, though – it had sharp teeth and claws, making it a fearsome beast.

Which one would you rather run into if they still roamed the Earth today?

# THE LIVES OF DINOSAURS TODAY

Both humans and dinosaurs would lead completely different lives if we existed alongside each other. Imagine walking along the beach and spotting ammonites in the shallow water, then seeing a pack of flying dinos whoosh overhead! What do you think would be the biggest change to your day-to-day life?

## SAVANNAH SEISMOSAUR

There has always been more plant-eating animals than meat-eating animals. On the African savanna today, you would see whole herds of Seismosaurus or Brachiosaurus, but only a handful of lions or cheetahs.

## MOUNTAIN HOMES

Short-needled conifers – a staple of the herbivore diet – aren't as plentiful as they were in the dinosaur times. They're mostly found in high **altitudes** and mountains. The fussiest long-necked plant-eaters, if unwilling or unable to change their diets, would be found thriving in the mountains today (if they could stand the cold!).

## PET DINOSAURS

Small, feathered dinosaurs, such as Archaeopteryx, would make good pets. They would probably need the same amount of care as exotic birds, such as parrots. It's thought they ate small reptiles, mammals and insects which means they would adapt easily to modern food. Would you want to look after a pet dinosaur?

## AMMONITE SUSHI

If smaller ocean animals were present today in the same numbers as in the Mesozoic Era, then they would make up a significant part of the ecosystem. If they were found to be edible to humans, and we were fast enough to catch them, then you'd be having dino and chips or ammonite sushi for dinner!

# DID YOU KNOW?

Dinosaurs are fascinating creatures. Scientists are constantly discovering more about them and finding answers to the world's most curious questions. Did you know these amazing facts about dinosaurs?

## HOW DO WE KNOW HOW OLD FOSSILS ARE?

Palaeontologists study the carbon in the bones, but they also look at the age of the rock that the fossil was extracted from. They do this by studying the chemicals and elements in the rock itself.

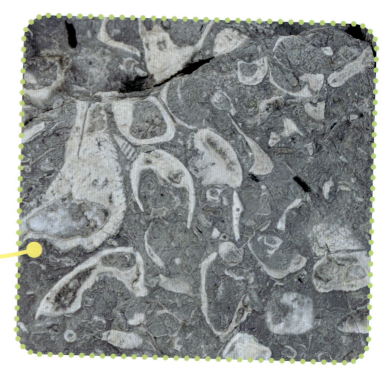

## HOW LONG DOES FOSSIL EXCAVATION TAKE?

Palaeontologists working on fossil sites can spend weeks or even months carefully studying the ground, collecting material, and brushing dirt away to unearth the fossils. They must be very careful not to damage the fossils, so work is slow.

## WERE ANY DINOSAURS RENOWNED FOR THEIR BRILLIANT CAMOUFLAGE?

It's extremely rare to find dinosaur skin, so it's a difficult question to answer. However, it seems likely that some dinosaurs would've learned to camouflage themselves - to either help them hide from predators, or sneak up on prey themselves! Many modern animals do this; why wouldn't dinosaurs have had similar habits?

## WHERE DINOSAURS WARM OR COLD-BLOODED?

For years, palaeontologists have struggled to work this out, however they now have an answer. Opposite to modern-day reptiles, most dinosaurs were warm-blooded. This means that they could regulate their own body temperature, explaining how they were able to live in different, extreme climates.

## WERE DINOSAUR BONES HEAVY?

Not as heavy as you'd think compared to their size! Many dino bones were hollow or had air sacs which made their skeletons lighter and allowed them to grow bigger, too.

## DID DINOSAURS ROAR?

Palaeontologists studying fossil vocal organs think dinosaurs were more likely to have growled with their mouths closed than roared with their mouths open! Some, like Olorotitan, might've made booming or honking noises.

# TRUE OR FALSE?

There are lots of fun facts about dinosaurs, but what is actually true and what is a myth? Put your dino knowledge to the test with this true or false quiz.

## WE COULD BRING DINOSAURS TO LIFE FROM THEIR FOSSILISED DNA!

FALSE. This makes a good story, but it's not good science. There's no way that DNA (the building blocks of all creatures) can survive intact over tens of millions of years.

## THERE ARE DINOSAURS ALIVE TODAY!

FALSE. Although some scientists think that birds should be called dinosaurs because they are so closely related. These scientists call the animals that we would regard as dinosaurs "non-avain dinosaurs" to distinguish them from birds which they call "**avian** dinosaurs".

## PLACES HAVE BEEN NAMED AFTER DINOSAURS!

TRUE. One town in Colorado (USA) has been renamed 'Dinosaur' due to its proximity to the Dinosaur National Monument! It's even got streets named after dinosaurs, such as Brontosaurus Boulevard and Triceratops Terrace.

## DINOSAURS ONLY LIVED IN HOT PLACES!

FALSE. Dinosaurs lived all over the world. We've discovered fossils of polar dinosaurs who would have lived in cold, snowy areas. Because Earth's **continents** are always moving, the climates where fossils are found were once different. For example, what is now Australia was within the Antarctic Circle when dinosaurs roamed there.

## ALL DINOSAURS HATCHED FROM EGGS!

TRUE. All evidence uncovered so far suggests that all dinosaurs reproduced by laying eggs, as most modern reptiles do. It's difficult, and sometimes impossible, to work out which dinosaurs laid the fossilised eggs that have been found, because only a few baby dinosaurs have ever been found inside them.

## MOST DINOSAURS HAD SCALES RATHER THAN SKIN!

TRUE. Fossilised skin is hardly ever found, but when it is, we can see that it usually consisted of flat scales – not overlapping like on many fish but adjoining, like a crocodiles. Sometimes there were bony lumps and horns set into the scales. We also know that some dinosaurs were covered in feathers.

# UNCOVERING THE PAST

Can we see live dinosaurs today? Yes, if we count birds as dinosaurs. No, if we're thinking about the big dinosaurs in this book. In this case, we must rely on fossil evidence to tell us what they looked like. Since the 1820s, when the first dinosaur fossils were found, we've been building up our knowledge piece by piece (literally).

Our knowledge of dinosaurs is far from complete: it's very rare for scientists to uncover fossils. Not only are fossils usually deep underground, but it's rare for land-living animals to fossilise in the first place. To fossilise, the animal needs to have been buried quickly, otherwise bones get scattered and rot away from exposure to weather and **bacteria**.

Archaeopteryx was first discovered in the 1860s and is still one of the oldest feathered dinosaurs known.

Fossils can be discovered in different circumstances. Usually, we only find isolated fossil bones which have been separated from the rest of the skeleton. They're not entirely useless - even an isolated tooth or limb bone may be identifiable to a species. However, because the bone has spent millions of years in the ground, it's often weathered or fragmented, making it tricky to find out which dinosaur it came from.

Cryptoclidus is one of the most studied and better understood plesiosaurs as palaeontologists have found many, mostly complete, fossils. The well-preserved fossils have shown their growth from very young to very old which gives us a good understanding of them.

Only very occasionally are articulated skeletons uncovered – this is when the bones are still joined together, as they were in life. More commonly, associated skeletons are found – this is when the bones are jumbled up, but it's obvious that they came from the same animal. It takes a very knowledgeable scientist to put the bones back together.

*Group of associated skeletons.*

## ANYONE CAN FIND A FOSSIL!

You don't need to be a professional palaeontologist to discover a prehistoric creature. Many have been found by children. In 2021, a young girl discovered a 220-million-year-old dinosaur footprint! Given that dinosaurs appeared 230 million years ago, this footprint must be from one of the earliest dinosaurs to walk the Earth.

Perhaps the greatest fossil hunter of all was 12-year-old Many Anning, who found and excavated the first complete skeleton of the prehistoric marine reptile **Ichthyosaurus** in England in 1811.

Why not begin your own searches by joining a fossil-hunting group?

*Statue of Mary Anning, fossil hunter in Lyme Regis, Dorset (England).*

# INDEX

# GLOSSARY

**Altitude** - The height of something compared to sea level.

**Arid** - Having little or no rain; dry and barren.

**Avian** - Relating or related to birds.

**Bacteria** - Microscopic organisms that can cause disease.

**Baleen** - The bony, flexible strips in the upper jaws of whales that feed by filtering food from the ocean water.

**Carnivore** - An animal which eats only meat.

**Conifer** - A type of tree bearing cones and needle- or scale-shaped leaves.

**Continent** - One of the world's large expanses of land.

**Controversial** – Relating to a topic on which people have opposite views.

**Excavate** (verb) - The careful removal of earth from an area in order to find buried remains.

**Fossil** - The remains or impression of a prehistoric plant or animal embedded in rock and preserved.

**Habitat** – The place or environment where a plant or animal naturally lives or grows.

**Herbivore** - An animal which eats only plants.

**Ichthyosaur** - A type of swimming reptile from the Mesozoic Era. They had streamlined fish-like bodies with fins and a tail.

**Mass extinction** - An event that brings about the extinction of a large number of animals and plants. There have been about five mass extinctions in the history of life on Earth.

**Meteorite** - A rock from space.

**Mesozoic Era** – The era of time in which dinosaurs lived, among other animals. It lasted around 186 million years, from 252 to 66 million years ago.

**Oases** - Green areas in a desert, where there is water and plants grow.

**Palaeontology** - The study of ancient life and fossils. People who study this are called palaeontologists.

**Period** - A division of geological time that can be defined by the types of animals or plants that existed then. Typically, a period lasts for tens of millions of years.

**Piscivore** - An animal which eats mostly fish.

**Plesiosaur** - A large marine reptile of the Mesozoic Era, with large, paddle-like limbs and a long, flexible neck.

**Sauropod** – A type of extremely large herbivorous dinosaur that had a long neck and tail, trunk-like legs, but small head.

**Scavenger** - An animal that feeds off food other animals have killed.

**Tsunami** – An extremely long and high fast-moving sea wave caused by an earthquake or other disturbance.